Climate Change

Impact on Agriculture and Costs of Adaptation

Gerald C. Nelson, Mark W. Rosegrant, Jawoo Koo, Richard Robertson, Timothy Sulser, Tingju Zhu, Claudia Ringler, Siwa Msangi, Amanda Palazzo, Miroslav Batka, Marilia Magalhaes, Rowena Valmonte-Santos, Mandy Ewing, and David Lee

International Food Policy Research Institute
Washington, D.C.

Updated October 2009

ABOUT IFPRI

The International Food Policy Research Institute (IFPRI) was established in 1975. IFPRI is one of 15 agricultural research centers that receives its principal funding from governments, private foundations, and international and regional organizations, most of which are members of the Consultative Group on International Agricultural Research.

Cover illustration adapted from photography by © Sven Torfinn/PANOS and © 2009 Klaus von Grebmer/IFPRI.

DOI: 10.2499/0896295354

ISBN 10-digit: 0-89629-535-4
ISBN 13-digit: 978-0-89629-535-3

Climate-Change Scenarios[3]

The research underlying this report provides detailed estimates of the impacts of climate change on agricultural production, consumption, prices, and trade, and also estimates the costs of adaptation. It uses a global agricultural supply-and-demand projection model (IMPACT 2009) linked to a biophysical crop model (DSSAT) of the impact of climate change on five important crops: rice, wheat, maize, soybeans, and groundnuts (see box). The report assesses climate-change effects on food security and human well-being using two indicators: per capita calorie consumption and child malnutrition numbers. It estimates the cost of investments—in three primary sources of increased agricultural productivity (agricultural research, rural roads, and irrigation)—needed to return the values of these two indicators from their 2050 values with climate change to their 2050 values without climate change. In other words, this report isolates the effects of climate change on future well-being and identifies only the costs of compensating for climate change.

IMPACT 2009

The IMPACT model was originally developed by the International Food Policy Research Institute (IFPRI) for projecting global food supply, food demand, and food security to 2020 and beyond.[4] It analyzes 32 crop and livestock commodities in 281 regions of the world that together cover the earth's land surface (with the exception of Antarctica). These regions are called food production units (FPUs). Production and demand relationships in countries are linked through international trade flows. The model simulates growth in crop production, determined by crop and input prices, externally determined rates of productivity growth and area expansion, investment in irrigation, and water availability. Demand is a function of prices, income, and population growth and contains four categories of commodity demand—food, feed, biofuels, and other uses. The 2009 version of the model includes a hydrology model and links to the Decision Support System for Agrotechnology Transfer (DSSAT) crop-simulation model, with yield effects of climate change at 0.5-degree intervals aggregated up to the food-production-unit level.

The DSSAT model is used to assess climate-change effects and CO_2 fertilization for five crops–rice, wheat, maize, soybeans, and groundnuts. For the remaining crops in IMPACT, the primary assumption is that plants with similar photosynthetic metabolic pathways will react similarly to any given climate-change effect in a particular geographic region. Millet, sorghum, sugarcane, and maize all follow the same (C4) metabolic pathway and are assumed to follow the DSSAT results for maize, in the respective geographic regions. The other crops in IMPACT follow a different pathway (C3), so the climate effects are assumed to follow the average for wheat, rice, soy, and groundnuts from the same geographic region, with two exceptions. The IMPACT commodities of "other grains" and dryland legumes are directly mapped to the DSSAT results for wheat and groundnuts, respectively.

Because climate-change simulations are inherently uncertain, two climate models have been used to simulate future climate, using the A2[5] scenario of the IPCC's Fourth Assessment Report: the National Center for Atmospheric Research, US (NCAR) model and the Commonwealth Scientific and Industrial Research Organization, Australia (CSIRO) model. We refer to the combination of model runs with A2 inputs as the NCAR and CSIRO scenarios. Both scenarios project higher temperatures in 2050, resulting in higher evaporation and increased precipitation as this water vapor returns to earth. The "wetter" NCAR scenario estimates average precipitation increases on land of about 10 percent, whereas the "drier" CSIRO scenario estimates increases of about 2 percent. Figure 1 shows the change in average maximum temperature between 2000 and 2050 for the CSIRO and NCAR scenarios. Figure 2 shows changes in average precipitation. In each set of figures, the legend colors are identical; a specific color represents the same change in temperature or precipitation across the two scenarios.

Figure 1—Change in average maximum temperature (°C), 2000–2050

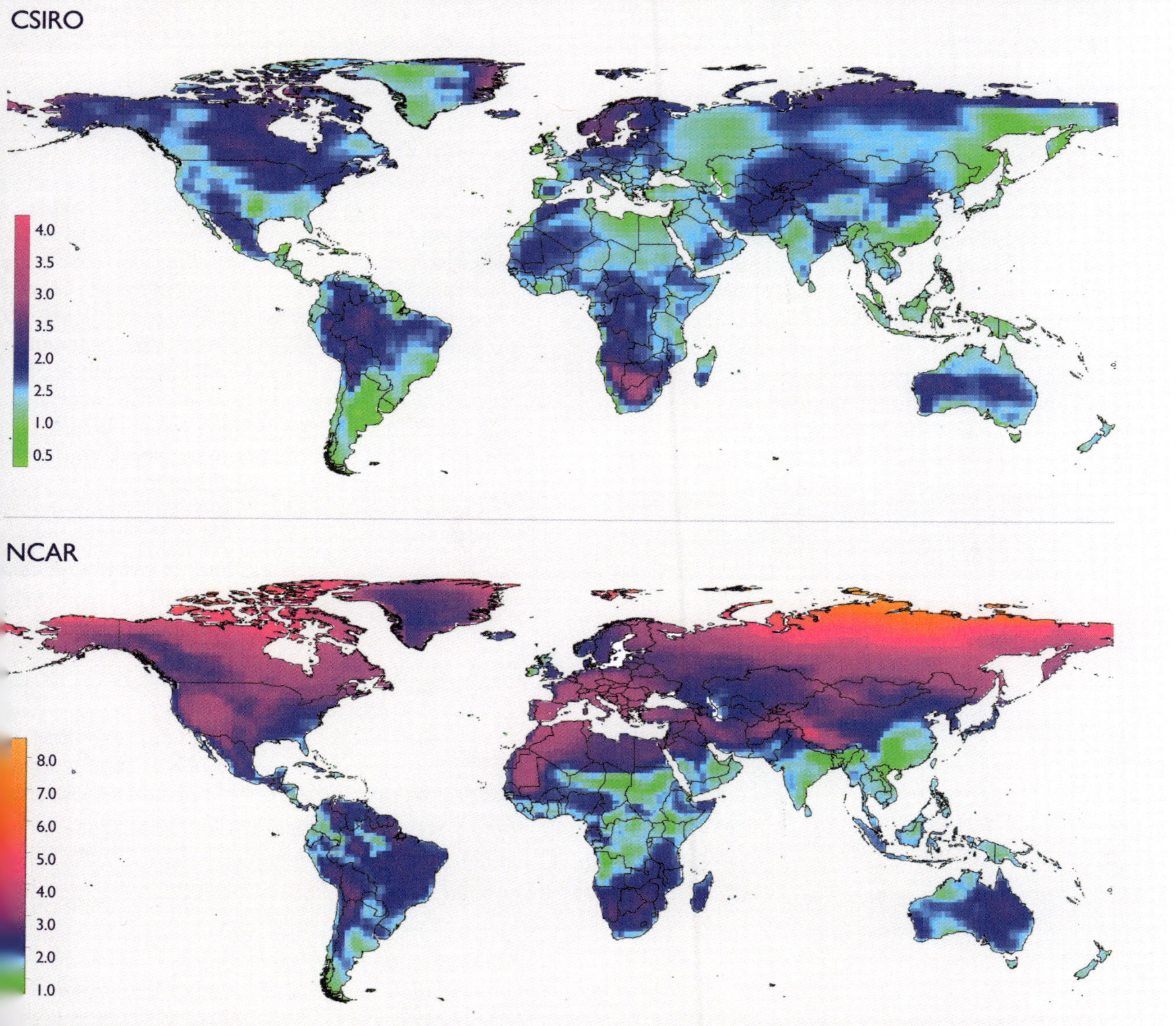

rce: Authors' calculations.

Figure 2—Change in precipitation (mm), 2000–2050

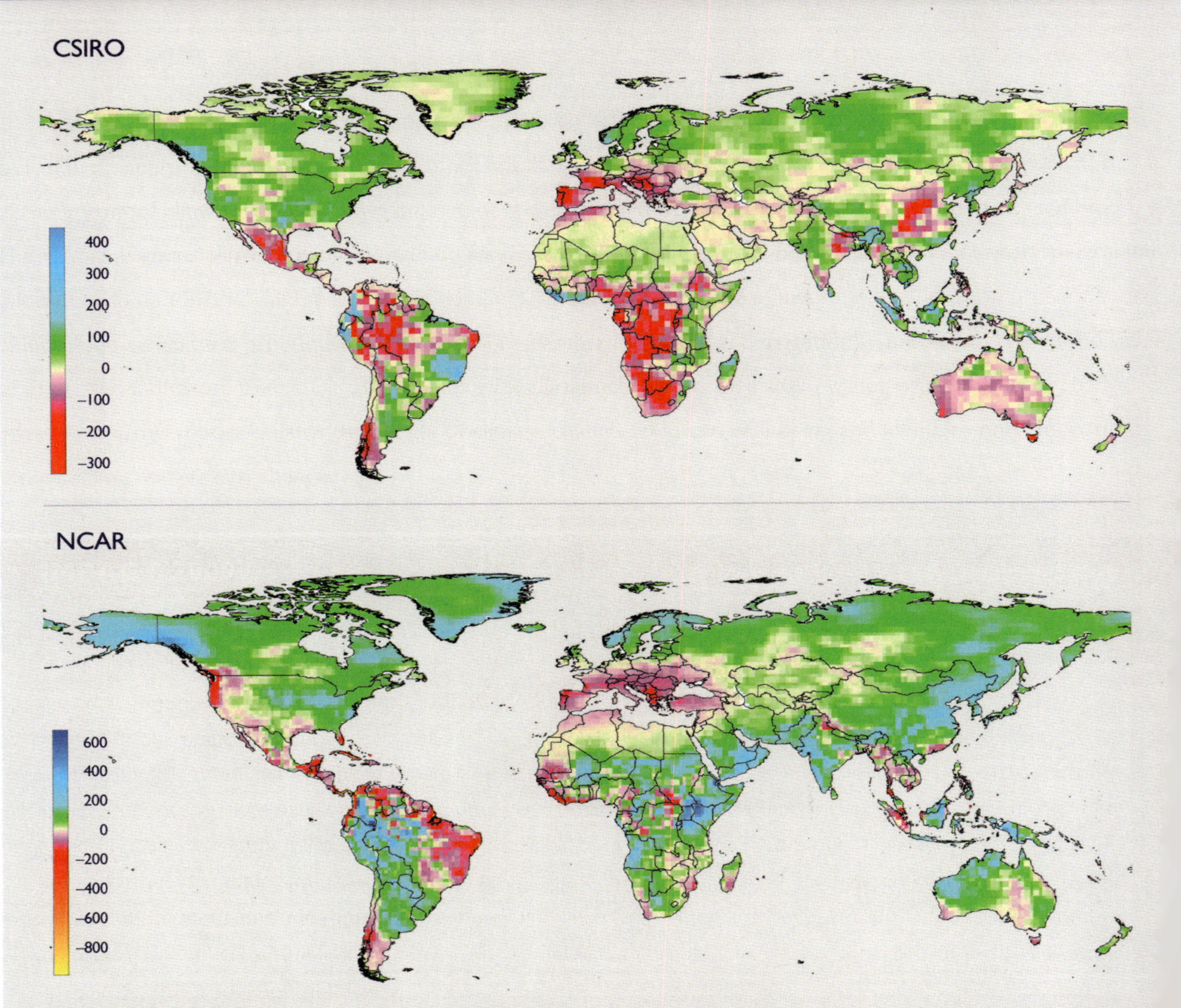

Source: Authors' calculations.

A quick glance at these figures shows that substantial differences exist across the two scenarios. For example, the NCAR scenario has substantially higher average maximum temperatures than does CSIRO. The CSIRO scenario has substantial precipitation declines in the western Amazon while NCAR shows declines in the eastern Amazon. The NCAR scenario has higher precipitation in Sub-Saharan Africa than d CSIRO. Northern China has both higher temperatu and more precipitation under NCAR than under CSIRO. These figures qualitatively illustrate the rang of potential climate outcomes using current modeli capabilities and provide an indication of the uncerta in climate-change impacts.

Impacts of Climate Change

The impacts of climate change on agriculture and human well-being include: 1) the biological effects on crop yields; 2) the resulting impacts on outcomes including prices, production, and consumption; and 3) the impacts on per capita calorie consumption and child malnutrition. The biophysical effects of climate change on agriculture induce changes in production and prices, which play out through the economic system as farmers and other market participants adjust autonomously, altering crop mix, input use, production, food demand, food consumption, and trade.

1. The Biological Effects of Climate Change on Yields

ising temperatures and changes in rainfall patterns ave direct effects on crop yields, as well as indirect ffects through changes in irrigation water availability.

irect effects on yields: rainfed and irrigated crops

ible 1 reports the direct biological effects of the o climate-change scenarios on crop yields modeled ectly with DSSAT for rainfed and irrigated crops in veloping and developed countries,[6] with and without O_2 fertilization (CF and No CF).[7] These results are ated by "growing" each crop around the world at -degree intervals with 2000 climate, growing them in with a 2050 scenario value, and then calculating ratio. In other words, no economic adjustments are uded. The rainfed yield changes are driven by both ipitation and temperature changes; the irrigated d effects are from temperature changes alone.

In developing countries, yield declines predominate nost crops without CO_2 fertilization. Irrigated at and irrigated rice are especially hard hit. On ige, yields in developed countries are affected less those in developing countries. For a few crops, te change actually increases developed-country . In calculating these projections, the East Asia and c region combines China, which is temperate for ost part, and Southeast Asia, which is tropical. ifferential effects of climate change in these two e zones are concealed. In China, some crops fare ably well because higher future temperatures orable in locations where current temperatures are at the low end of the crop's optimal temperature. Yields of important crops in Southeast Asia fall substantially in both scenarios unless CO_2 fertilization is effective in farmers' fields.

South Asia is particularly hard hit by climate change. For almost all crops, it is the region with the greatest yield decline. With CO_2 fertilization, the yield declines are lower; in many locations, some yield increases occur relative to 2000. However, rainfed maize and irrigated and rainfed wheat still see substantial areas of reduced yields. Sub-Saharan Africa sees mixed results, with small declines or increases in maize yields and large negative effects on rainfed wheat. The Latin America and Caribbean region also has mixed yield effects, with some crops up slightly and some down.

Indirect effects: Irrigated crops

Climate change will have a direct impact on water availability for irrigated crops. Internal renewable water (IRW) is the water available from precipitation. Both climate scenarios result in more precipitation over land than would occur with no climate change. Under the NCAR scenario, all regions experience increased IRW. Under the CSIRO scenario, the average IRW increase is less than occurs with NCAR, and the Middle East and North Africa and Sub-Saharan Africa regions both experience reductions of about 4 percent.

In addition to precipitation changes, climate change-induced higher temperatures increase the water requirements of crops. The ratio of water consumption to requirements is called irrigation water supply reliability (IWSR). The smaller the ratio, the greater the water stress on irrigated crop yields.

Table 1—Climate-change induced yield effects by crop and management system, % change from yield with 2000 climate to yield with 2050 climate

Region	CSIRO No CF	NCAR No CF	CSIRO CF	NCAR CF
Maize, irrigated				
Developing countries	–2.0	–2.8	–1.4	–2.1
Developed countries	–1.2	–8.7	–1.2	–8.6
Maize, rainfed				
Developing countries	0.2	–2.9	2.6	–0.8
Developed countries	0.6	–5.7	9.5	2.5
Rice, irrigated				
Developing countries	–14.4	–18.5	2.4	–0.5
Developed countries	–3.5	–5.5	10.5	9.0
Rice, rainfed				
Developing countries	–1.3	–1.4	6.5	6.4
Developed countries	17.3	10.3	23.4	17.8
Wheat, irrigated				
Developing countries	–28.3	–34.3	–20.8	–27.2
Developed countries	–5.7	–4.9	–1.3	–0.1
Wheat, rainfed				
Developing countries	–1.4	–1.1	9.3	8.5
Developed countries	3.1	2.4	9.7	9.5

Source: Compiled by authors.
Note: For each crop and management system, this table reports the area weighted average change in yield for a crop grown with 2050 climate instead of 2000 climate. CF = with CO_2 fertilization; No CF = without CO_2 fertilization.

Across the group of developing countries, IWSR improves under the NCAR scenario and worsens under the CSIRO scenario. However, regional differentiation of climate-change effects is important. IWSR improves slightly for the Latin America and Caribbean region and for the Middle East and North Africa, but worsens slightly for Sub-Saharan Africa under both scenarios. For East Asia and the Pacific and for South Asia, reliability increases under the NCAR scenario but declines under the CSIRO scenario.

Yield reductions of irrigated crops due to water stress are directly estimated in the hydrology portion of IMPACT, taking into account the growing demand for water outside agriculture as well as agricultural demands. As expected, irrigated yield losses due to water stress are relatively higher under the CSIRO scenario than the NCAR scenario. For example, in East Asia and the Pacific, with no climate change, the combined effects of nonagricultural demand growth and increased irrigated area result in an average 4.8-percent decline in irrigated rice yields. Under the NCAR scenario, that decline is only 1.2 percent However, under the drier CSIRO scenario, the irrigated yield loss from water stress is 6.7 percent In East Asia and the Pacific, irrigated rice, wheat, an maize yield losses are all large under the CSIRO model. South Asia irrigated yields for all crops wou experience large declines under both scenarios. In Sub-Saharan Africa, maize yields are less under bot models, but the CSIRO effects are especially large. Latin America and the Caribbean yields are relativ unaffected, in part due to the small amount of irrigated production in that region.

2. Prices, Production, and Food Consumption

Prices

World prices are a useful single indicator of the effects of climate change on agriculture. Table 2 reports the effects of the two climate-change scenarios on world food prices, with and without CO_2 fertilization. It also reports the effects with no climate change. Figures 3 and 4 demonstrate world price effects for livestock production and major grains, respectively, assuming no CO_2 fertilization.

With no climate change, world prices for the nost important agricultural crops—rice, wheat, maize, nd soybeans will increase between 2000 and 2050, riven by population and income growth and biofuels emand. Even with no climate change, the price of ce would rise by 62 percent, maize by 63 percent, ybeans by 72 percent, and wheat by 39 percent. limate change results in additional price increases— total of 32 to 37 percent for rice, 52 to 55 percent r maize, 94 to 111 percent for wheat, and 11 to percent for soybeans. If CO_2 fertilization is ective in farmers' fields, these 2050 prices are percent smaller.

Livestock are not directly affected by climate nge in the IMPACT model, but the effects of higher d prices caused by climate change pass through to stock, resulting in higher meat prices. For example, f prices are 33 percent higher by 2050 with no ate change and 60 percent higher with climate nge and no CO_2 fertilization of crops. With CO_2 lization, crop-price increases are less, so the beef- e increase is about 1.5 percent less than with no fertilization.

luction

3 reports the effects of climate change on production in 2050 compared to production ut climate change, based on the NCAR and O scenarios, accounting for both the direct es in yield and area caused by climate change and omous adaptation as farmers respond to changing with changes in crop mix and input use. The ve effects of climate change on crop production pecially pronounced in Sub-Saharan Africa and Asia. In South Asia, the climate scenario results -percent decline in rice production relative to the no–climate-change scenario, a 44- to 49-percent decline in wheat production, and a 9- to 19-percent fall in maize production. In Sub-Saharan Africa, the rice, wheat, and maize yield declines with climate change are 15 percent, 34 percent, and 10 percent, respectively. For East Asia and the Pacific, the results are mixed and depend on both the crop and the model used. Rice production declines by around 10 percent, wheat production increases slightly, and maize production declines with the drier CSIRO scenario but increases with the NCAR scenario. Comparing average production changes, developing countries fare worse for all crops under both the CSIRO and NCAR scenarios than do developed countries.

Food Consumption

Agricultural output used for human consumption is determined by the interaction of supply, demand, and the resulting prices with individual preferences and income. Table 4 shows average per capita consumption of cereals and meat products in 2000 and in 2050 under the CSIRO and NCAR models, with and without CO_2 fertilization. It also reports consumption with no climate change.

Without climate change, rising per capita income results in reduced declines in per capita consumption of cereals in developing countries between 2000 and 2050 and increased meat consumption increases, with the meat increases more than offsetting the decline in cereals. Climate change reduces the growth in meat consumption slightly and causes a more substantial fall in the consumption of cereals. These results are the first indication of the negative welfare effects due to climate change. Both models have similar effects.

3. Per Capita Calorie Consumption and Child Malnutrition

The primary measures used for the effects of climate change on human welfare are the change in calorie availability and the change in the number of malnourished children between 2000 and 2050 without climate change, and in 2050 using the two climate-change scenarios.

The declining consumption of cereals translates into similarly large declines in calorie availability as the result of climate change (see Figure 5 and Tables 5 and 6). Without climate change, calorie availability increases

Table 2—World food prices (US$/metric ton) in 2000 and 2050 and percent changes for selected crops and livestock products

Agricultural product	2000	2050 No climate change	2050 NCAR no CF	2050 CSIRO no CF	2050 NCAR CF effect (% change from no CF)	2050 CSIRO CF effect (% change from no CF)
Rice (US$/mt)	190	307	421	406	–17.0	–15.1
% change from 2000		61.6	121.2	113.4		
% change from 2050, no climate change			36.8	32.0		
Wheat (US$/mt)	113	158	334	307	–11.4	–12.5
% change from 2000		39.3	194.4	170.6		
% change from 2050, no climate change			111.3	94.2		
Maize (US$/mt)	95	155	235	240	–11.2	–12.6
% change from 2000		63.3	148.0	153.3		
% change from 2050, no climate change			51.9	55.1		
Soybeans (US$/mt)	206	354	394	404	–60.6	–62.2
% change from 2000		72.1	91.6	96.4		
% change from 2050, no climate change			11.4	14.2		
Beef (US$/mt)	1,925	2,556	3,078	3,073	–1.3	–1.5
% change from 2000		32.8	59.8	59.6		
% change from 2050, no climate change			20.4	20.2		
Pork (US$/mt)	911	1,240	1,457	1,458	–1.3	–1.5
% change from 2000		36.1	60.0	60.1		
% change from 2050, no climate change			17.5	17.6		
Lamb (US$/mt)	2,713	3,102	3,462	3,461	–0.7	–0.8
% change from 2000		14.4	27.6	27.6		
% change from 2050, no climate change			11.6	11.6		
Poultry (US$/mt)	1,203	1,621	1,968	1,969	–1.9	–2.
% change from 2000		34.7	63.6	63.6		
% change from 2050, no climate change			21.4	21.5		

Source: Compiled by authors.
Note: Prices are in 2000 US$.

Figure 3—World prices, Livestock products

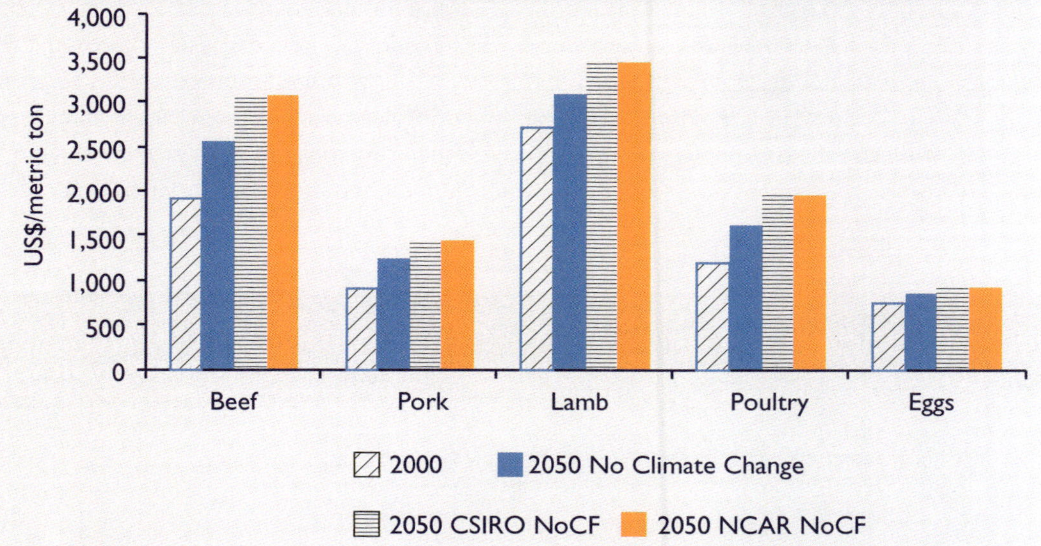

Source: Compiled by authors.
Note: Prices are in 2000 US$.

Figure 4—World prices, Major grains

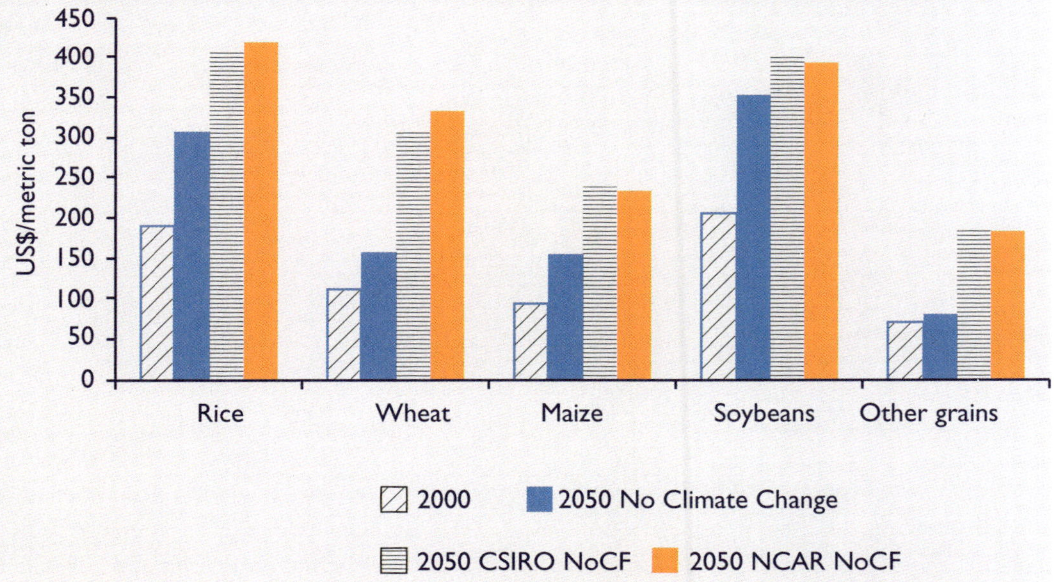

Source: Compiled by authors.
Note: Prices are in 2000 US$.

Table 3—Climate-change effects on crop production, no CO_2 fertilization

Agricultural product	South Asia	East Asia and the Pacific	Europe and Central Asia	Latin America and the Caribbean	Middle East and North Africa	Sub-Saharan Africa	Developed countries	Developing countries	Wor
Rice									
2000 (mmt)	119.8	221.7	1.1	14.8	5.5	7.4	20.4	370.3	390.
2050 No CC (mmt)	168.9	217.0	2.6	17.8	10.3	18.3	20.3	434.9	455.
2050 No CC (% change)	41.0	–2.1	144.4	19.8	87.4	146.0	–0.3	17.4	16.
CSIRO (% change)	–14.3	–8.1	–0.2	–21.7	–32.9	–14.5	–11.8	–11.9	–11
NCAR (% change)	–14.5	–11.3	–0.8	–19.2	–39.7	–15.2	–10.6	–13.6	–13
Wheat									
2000 (mmt)	96.7	102.1	127.5	23.5	23.6	4.5	205.2	377.9	583
2050 No CC (mmt)	191.3	104.3	252.6	42.1	62.0	11.4	253.7	663.6	91
2050 No CC (% change)	97.9	2.1	98.1	78.7	162.3	154.4	23.6	75.6	5
CSIRO (% change)	–43.7	1.8	–43.4	11.4	–5.1	–33.5	–7.6	–29.2	–2
NCAR (% change)	–48.8	1.8	–51.0	17.4	–8.7	–35.8	–11.2	–33.5	–2
Maize									
2000 (mmt)	16.2	141.8	38.0	80.1	8.2	37.1	297.9	321.3	6
2050 No CC (mmt)	18.7	264.7	62.7	143.1	13.1	53.9	505.1	556.2	1.06
2050 No CC (% change)	15.7	86.6	65.1	78.8	59.4	45.3	69.6	73.1	[illegible]
CSIRO (% change)	–18.5	–12.7	–19.0	–0.3	–6.8	–9.6	11.5	–10.0	[illegible]
NCAR (% change)	–8.9	8.9	–38.3	–4.0	–9.8	–7.1	1.8	–2.3	[illegible]
Millet									
2000 (mmt)	10.5	2.3	1.2	0.0	0.0	13.1	0.5	27.3	[illegible]
2050 No CC (mmt)	12.3	3.5	2.1	0.1	0.1	48.1	0.8	66.2	[illegible]
2050 No CC (% change)	16.5	50.1	77.2	113.0	128.0	267.2	60.5	142.5	[illegible]
CSIRO (% change)	–19.0	4.2	–4.3	8.8	–5.5	–6.9	–3.0	–8.5	[illegible]
NCAR (% change)	–9.5	8.3	–5.2	7.2	–2.7	–7.6	–5.6	–7.0	[illegible]
Sorghum									
2000 (mmt)	8.4	3.1	0.1	11.4	1.0	19.0	16.9	43.0	[illegible]
2050 No CC (mmt)	9.6	3.4	0.4	28.0	1.1	60.1	20.9	102.6	[illegible]
2050 No CC (% change)	13.9	11.6	180.9	145.3	12.2	216.9	23.6	138.7	[illegible]
CSIRO (% change)	–19.6	1.4	–2.7	2.3	0.3	–2.3	–3.1	–2.5	[illegible]
NCAR (% change)	–12.2	6.7	–10.4	4.3	0.7	–3.0	–7.3	–1.5	[illegible]

Source: Compiled by authors.
Note: The rows labeled "2050 No CC (% change)" indicate the percent change between production in 2000 and 2050 with no climate change. The rows labeled change)" and "NCAR (% change)" indicate the additional percent change in production in 2050 due to climate change relative to 2050 with no climate change. Fo South Asia sorghum production was 8.4 mmt in 2000. With no climate change, South Asia sorghum production is predicted to increase to 9.6 mmt in 2050, an ir 13.9 percent. With the CSIRO scenario, South Asia sorghum production in 2050 is 19.6 percent lower than with no climate change in 2050 (7.72 mmt instead of mmt = million metric tons.

throughout the world between 2000 and 2050. The largest increase, of 13.8 percent, is in East Asia and the Pacific, but there are gains for the average consumer in all countries—by 3.7 percent in Latin America, 5.9 percent in Sub-Saharan Africa, and 9.7 percent in South Asia.

With climate change, however, calorie availability in 2050 is not only lower than the no-climate-change scenario in 2050—it actually declines relative to 2000 levels throughout the world. For the average consumer in a developing country, the decline is 10 percent relative to 2000. With CO_2 fertilization, the declines are 3 percent to 7 percent less severe, but are still large relative to the no–climate-change scenario. There is almost no difference in calorie outcome between the two climate scenarios.

Table 4—Per capita consumption (kg per year) of cereals and meats with and without climate change (NCAR and CSIRO)

		2050				
Region	2000	No climate change	CSIRO no CF	NCAR no CF	CSIRO CF effect (% change relative to CSIRO no CF in 2050)	NCAR CF effect (% change relative to NCAR no CF in 2050)
Meat						
South Asia	6	16	14	14	0.9	0.8
East Asia and the Pacific	40	71	66	66	0.7	0.6
Europe and Central Asia	42	56	51	51	0.8	0.7
Latin America and the Caribbean	57	71	64	64	1.0	0.9
Middle East and North Africa	23	39	36	36	0.7	0.6
Sub-Saharan Africa	11	18	16	16	1.0	0.8
Developed countries	88	100	92	92	0.8	0.7
Developing countries	28	41	37	37	0.8	0.7
Cereals						
South Asia	164	157	124	121	7.0	7.1
East Asia and the Pacific	184	158	124	120	8.1	8.3
Europe and Central Asia	162	169	132	128	5.3	4.9
Latin America and the Caribbean	123	109	89	87	6.1	5.9
Middle East and North Africa	216	217	172	167	5.5	5.1
Sub-Saharan Africa	117	115	89	89	7.4	7.1
Developed countries	118	130	97	94	6.8	6.3
Developing countries	164	148	116	114	7.1	7.1

rce: Compiled by authors.

Figure 5—Daily per capita calorie availability with and without climate change

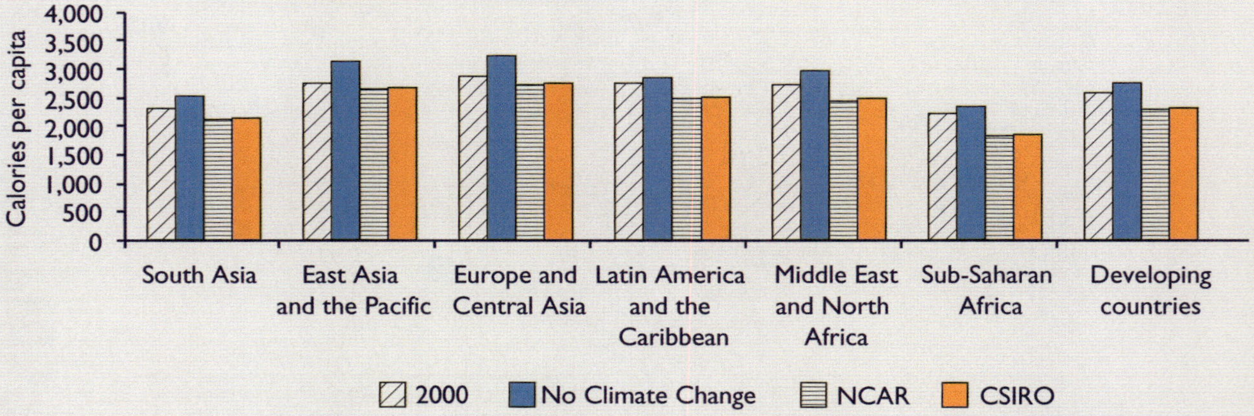

Source: Compiled by authors.

Table 5—Daily per capita calorie availability with and without climate change

Region	2000	2050 No climate change kcal/day	2050 NCAR no CF kcal/day	2050 CSIRO no CF kcal/day	2050 NCAR CF effects (% change relative to NCAR no CF in 2050)	2050 CSIRO CF effects (% change relative to CSIRO no CF in 2050)
South Asia	2,424	2,660	2,226	2,255	4.3	4.3
East Asia and the Pacific	2,879	3,277	2,789	2,814	4.3	4.3
Europe and Central Asia	3,017	3,382	2,852	2,885	2.7	2.9
Latin America and the Caribbean	2,879	2,985	2,615	2,628	2.7	2.8
Middle East and North Africa	2,846	3,119	2,561	2,596	3.6	3.7
Sub-Saharan Africa	2,316	2,452	1,924	1,931	6.5	6.9
Developed countries	3,450	3,645	3,190	3,215	2.3	2.5
Developing countries	2,696	2,886	2,410	2,432	4.4	4.4

Source: Compiled by authors.

Table 6—Total number of malnourished children in 2000 and 2050 (million children under 5 years of age)

Region	2000	2050				
		No climate change	NCAR no CF	CSIRO no CF	NCAR CF effects (% change relative to NCAR no CF in 2050)	CSIRO CF effects (% change relative to CSIRO no CF in 2050)
South Asia	76	52	59	59	–3	–3
East Asia and the Pacific	24	10	15	14	–9	–9
Europe and Central Asia	4	3	4	4	–4	–5
Latin America and and the Caribbean	8	5	6	6	–5	–5
Middle East and North Africa	3	1	2	2	–10	–11
Sub-Saharan Africa	33	42	52	52	–5	–6
All developing countries	148	113	139	137	–5	–5

Source: Compiled by authors.

Note: The last two columns in this table report the percentage difference between the number of malnourished children in 2050 with and without CO_2 fertilization. For example, under the NCAR model, assuming CO_2 fertilization is effective in the field, there would be a 3-percent decline in the number of malnourished children in South Asia relative to the climate change outcome without CO_2 fertilization.

Costs of Adaptation

Climate-change adaptation is increasingly on the agenda of researchers, policymakers, and program developers who are aware that climate change is real and threatens to undermine social and ecological sustainability. In agriculture, adaptation efforts focus on implementing measures that help build rural livelihoods that are more resilient to climate variability and disaster. This section provides an assessment of the costs of productivity-enhancing investments in agricultural research, rural roads, and irrigation infrastructure and efficiency that can help farmers adapt to climate change. First, regardless of climate-change scenario, agriculture will be negatively affected by climate change.

Climate change increases child malnutrition and reduces calorie consumption dramatically. Thus, aggressive agricultural productivity investments are needed to raise calorie consumption enough to offset the negative impacts of climate-change on the health and well-being of children.

In order to assess the costs of adaptation alone, it is important to identify agricultural productivity investments that reduce child malnutrition with climate change to no-climate-change levels, holding all other macro changes constant, such as income and population growth. Two scenarios are assessed. The first, shown in Table 7, focuses on developing countries and describes the investments needed to reduce childhood malnutrition close to level it would be without climatechange. The cost estimates are based only on productivity-enhancing investments in developing countries. The second experiment involves including additional productivity enhancements in developed countries to assess the potential for spillovers in the developing world.

Table 8 reports the effects on daily per capita calorie availability for these two scenarios. Table 9 reports the results for child malnutrition for the two climate models relative to the no–climate-change scenario. Figures 6 and 7 are graphs of the malnutrition counts for the various developing-country regions before and after the productivity-enhancing investments. Finally, Table 10 reports the annualized additional investment costs needed to counteract the effects of climate change on children.

Table 7—Developing-country agricultural productivity investments

60-percent increase in crop (all crops) yield growth over baseline
30-percent increase in animal numbers growth
40-percent increase in production growth of oils and meals
25-percent increase in irrigated area growth
15-percent decrease in rainfed area growth
15-percent increase in basin water efficiency by 2050

Source: Compiled by authors.

Table 8—Daily calorie per capita consumption with adaptive investments (kcals/person/day)

Scenario	South Asia	East Asia and the Pacific	Europe and Central Asia	Latin America and the Caribbean	Middle East and North Africa	Sub-Saharan Africa	Developing countries
2000	2,424	2,879	3,017	2,879	2,846	2,316	2,696
2050							
No climate change	2,660	3,277	3,382	2,985	3,119	2,452	2,886
NCAR	2,226	2,789	2,852	2,615	2,561	1,924	2,410
NCAR +	2,531	3,161	3,197	2,994	2,905	2,331	2,768
NCAR + +	2,564	3,198	3,235	3,027	2,941	2,367	2,803
CSIRO	2,255	2,814	2,885	2,628	2,596	1,931	2,432
CSIRO +	2,574	3,200	3,243	3,011	2,954	2,344	2,801
CSIRO ++	2,612	3,241	3,285	3,048	2,996	2,384	2,840

Source: Compiled by authors.
Note: NCAR + and CSIRO + include only agricultural productivity investments in the developing world. NCAR ++ and CSIRO ++ include all productivity improvements in both developing and developed countries. The climate change results presented in this table assume no CO_2 fertilization effects.

Table 9—Child malnutrition counts with adaptive investments (million children)

Scenario	South Asia	East Asia and the Pacific	Europe and Central Asia	Latin America and the Caribbean	Middle East and North Africa	Sub-Saharan Africa	Developing countries
2000	75.62	23.81	4.11	7.69	3.46	32.67	147.84
2050							
No climate change	52.29	10.09	2.70	4.98	1.10	41.72	113.33
NCAR	59.06	14.52	3.73	6.43	2.09	52.21	138.52
NCAR +	54.16	10.82	3.04	4.94	1.37	44.09	118.87
NCAR ++	53.66	10.48	2.97	4.83	1.32	43.47	117.18
CSIRO	58.56	14.25	3.66	6.37	2.01	52.06	137.39
CSIRO +	53.51	10.44	2.95	4.88	1.29	43.87	117.40
CSIRO ++	52.96	10.18	2.87	4.76	1.23	43.17	115.62

Source: Compiled by authors.
Note: NCAR + and CSIRO + include only agricultural productivity investments in the developing world. NCAR ++ and CSIRO ++ include all productivity improvements in both developing and developed countries. The climate change results presented in this table assume no CO_2 fertilization effects.

Figure 6—Child malnutrition effects, South Asia and Sub-Saharan Africa

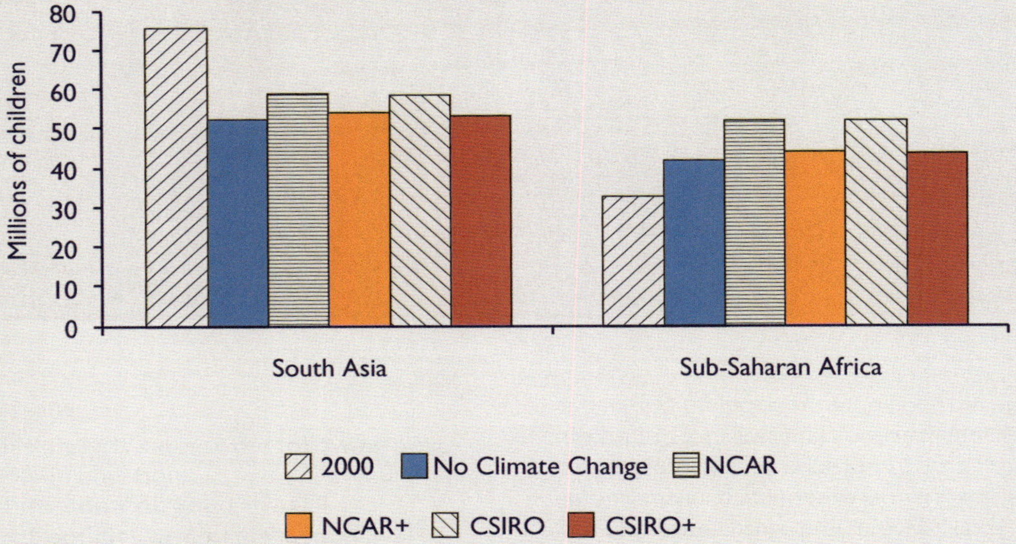

Source: Compiled by authors.

Figure 7—Child malnutrition effects, East Asia and the Pacific, Europe and Central Asia, Latin America and the Caribbean, and Middle East and North Africa

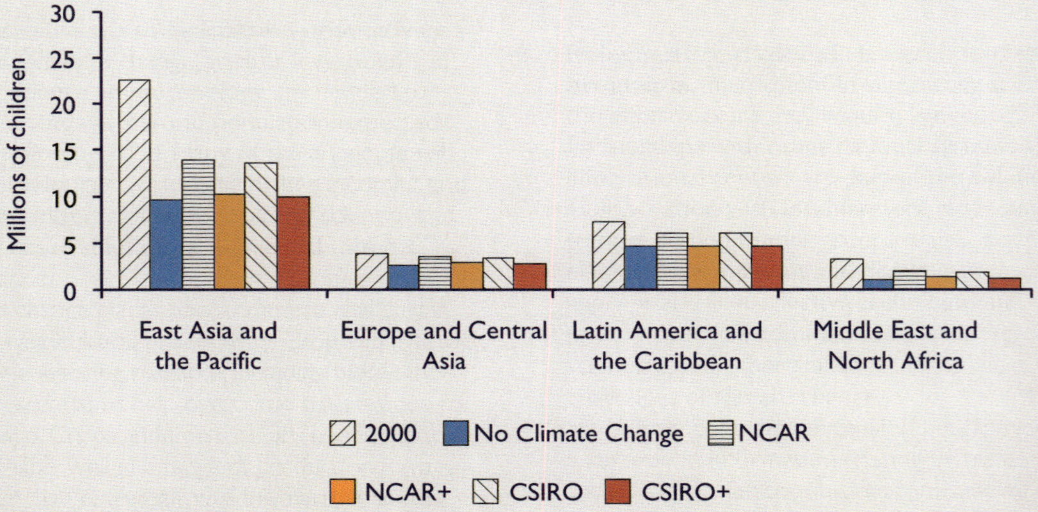

Source: Compiled by authors.

As shown in Table 10, the additional annual investments needed to return the child malnutrition numbers to the no climate-change results are $7.1 billion under the wetter NCAR scenario and $7.3 billion under the drier CSIRO scenario. Sub-Saharan African investment needs dominate, making up about 40 percent of the total. Of that amount, the vast majority is for rural roads. South Asia investments are about $1.5 billion per year, with Latin America and the Caribbean close behind with about $1.2 to $1.3 billion per year. East Asia and the Pacific needs are just under $1 billion per year. Agricultural research is important in all three of these regions, as are irrigation investments. Unlike Sub-Saharan Africa, road investments in these regions are relatively small.

With additional investments in developed countries, spillover effects to the developing world reduce the need for adaptation investments slightly. For example, with the NCAR scenario, the annual investment need is $7.1 billion if productivity expenditures are only in the developing world. With developed-country productivity investments, that amount drops to $6.8 billion.

The key messages embodied in these results point to the importance of improving the productivity of agriculture as a means of meeting the future challenges that climate change represents. The path to the needed agricultural productivity gains varies by region and to some extent, by climate scenario.

Table 10—Additional annual investment expenditure needed to counteract the effects of climate change on nutrition (million 2000 US$)

Scenario	South Asia	East Asia and the Pacific	Europe and Central Asia	Latin America and the Caribbean	Middle East and North Africa	Sub-Saharan Africa	Developing countries
NCAR with developing-country investments							
Agricultural research	172	151	84	426	169	314	1,316
Irrigation expansion	344	15	6	31	–26	537	907
Irrigation efficiency	999	686	99	129	59	187	2,158
Rural roads (area expansion)	8	73	0	573	37	1,980	2,671
Rural roads (yield increase)	9	9	10	3	1	35	66
Total	1,531	934	198	1,162	241	3,053	7,118
CSIRO with developing-country investments							
Agricultural research	185	172	110	392	190	326	1,373
Irrigation expansion	344	1	1	30	–22	529	882
Irrigation efficiency	1,006	648	101	128	58	186	2,128
Rural Roads (area expansion)	16	147	0	763	44	1,911	2,881
Rural Roads (yield increase)	13	9	11	3	1	36	74
Total	1,565	977	222	1,315	271	2,987	7,338

Source: Compiled by authors.
Note: These results are based on crop model yield changes that do not include the CO_2 fertilization effect.

Conclusion

This analysis brings together for the first time detailed modeling of crop growth under climate change with insights from an extremely detailed global agriculture model. The results show that agriculture and human well-being will be negatively affected by climate change. Crop yields will decline, production will be affected, crop and meat prices will increase, and consumption of cereals will fall, leading to reduced calorie intake and increased child malnutrition.

These stark results suggest the following policy and program recommendations:

- Design and implement good overall development policies and programs.
- Increase investments in agricultural productivity.
- Reinvigorate national research and extension programs.
- Improve global data collection, dissemination, and analysis.
- Make agricultural adaptation a key agenda point within the international climate negotiation process.
- Recognize that enhanced food security and climate-change adaptation go hand in hand.
- Support community-based adaptation strategies.
- Increase funding for adaptation programs by at least an additional $7 billion per year.

These investments may not guarantee that all the negative consequences of climate change can be overcome. But continuing with a "business-as-usual" approach will almost certainly guarantee disastrous consequences.

Notes

1. World Bank 2008.

2. All dollars are 2000 US dollars unless otherwise indicated.

3. For a full description of the methodology, see Appendix 1 (www.ifpri.org/sites/default/files/publications/pr21app1.pdf).

4. Rosegrant et al. 2008.

5. See Appendix 1 (www.ifpri.org/sites/default/files/publications/pr21app1.pdf) for description of A2 scenario.

6. To see the results for the full World Bank regional grouping of countries, see Table A2.1 in Appendix 2 (www.ifpri.org/sites/default/files/publications/pr21app2.pdf).

   Plants produce more vegetative matter as atmospheric concentrations of CO_2 increase. The effect depends on the nature of the photosynthetic process used by the plant species. Because the effects of higher concentrations of CO_2 on farmer's fields are uncertain, we report results both with 369 parts per million of atmospheric CO_2—the approximate concentration in 2000 (No CF results)—and 532 parts per million (CF results), the expected concentration in 2050 under the A2 scenario.

References

Fan, S., P. Hazell, and S. Thorat. 1998. *Government spending, growth and poverty: An analysis of interlinkages in rural India.* Environment and Production Technology Division Discussion Paper 33. Washington, D.C.: International Food Policy Research Institute.

Haie, N., and A. A. Keller. 2008. Effective efficiency as a tool for sustainable water resources management. *Journal of the American Water Resources Association* 10: 1752–1688.

IPCC et al. 2007. *Climate change 2007: Impacts, adaptation and vulnerability.* Contribution of Working Group II to the Fourth Assessment Report of the Intergovernmental Panel on Climate Change. Cambridge, UK: Cambridge University Press.

Jones, J. W., G. Hoogenboom, C. H. Porter, K. J. Boote, W. D. Batchelor, L. A. Hunt, P. W. Wilkens, U. Singh, A. J. Gijsman, and J. T. Ritchie. The DSSAT cropping system model. 2003. *European Journal of Agronomy* 18(3-4): 235–265.

Keller, A., and J. Keller. 1995. Effective efficiency: A water use concept for allocating freshwater resources. Winrock International, Center for Economic Policy Studies, Discussion Paper 22. Arlington, Va., U.S.A.: Winrock International.

Long, S. P., E. A. Ainsworth, A. D. B. Leakey, J. Nosberger, and D. R. Ort. 2006. Food for thought: Lower-than-expected crop yield stimulation with rising CO_2 concentrations. *Science* 312(5782): 1918–1921.

Parry, M. L., C. Rosenzweig, A. Iglesias, M. Livermore, and G. Fischer. 2004. Effects of climate change on global food production under SRES emissions and socio-economic scenarios. *Global Environmental Change* 14(1): 53–67.

Rosegrant, M. W., S. Msangi, C. Ringler, T. B. Sulser, T. Zhu, and S. A. Cline. 2008. *International Model for Policy Analysis of Agricultural Commodities and Trade (IMPACT): Model description.* Washington, D.C.: International Food Policy Research Institute.

Smith, L., and L. Haddad. 2000. *Explaining child malnutrition in developing countries: A cross-country analysis.* IFPRI Research Report. Washington, D.C.: International Food Policy Research Institute.

World Bank. 2008. World Development Report 2008: *Agriculture for Development.* Washington, D.C.: The World Bank.

You, L., and S. Wood. 2006. An entropy approach to spatial disaggregation of agricultural production. *Agricultural Systems* 90(1-3): 329–347.

Zavala, J. A., C. L. Casteel, E. H. DeLucia, and M. R. Berenbaum. 2008. Anthropogenic increase in carbon dioxide compromises plant defense again[st] invasive insects. *Proceedings of the National Acade[my] of Sciences* 105(13): 5129–5133.

About the Authors

All authors are part of IFPRI's Environment and Production Technology Division (EPTD). **Gerald C. Nelson** is senior research fellow, **Mark W. Rosegrant** is the division director, **Jawoo Koo** is a research fellow, **Richard Robertson** is a research fellow, **Timothy Sulser** is a scientist, **Tingju Zhu** is a senior scientist, **Claudia Ringler** is a senior research fellow, **Siwa Msangi** is a senior research fellow, **Amanda Palazzo** is a senior research assistant, **Miroslav Batka** is a research assistant, **Marialia Magalhaes** is a senior researc[h] assistant, **Rowena Valmonte-Santos** is a senior research analyst, **Mandy Ewing** is a research analyst, and **David Lee** is a consultant.

Printed in Dunstable, United Kingdom

91222394R00018